The Artemis Accords

The Artemis Accords

PRINCIPLES FOR COOPERATION IN THE CIVIL EXPLORATION, AND USE OF THE MOON, MARS, COMETS, AND ASTEROIDS FOR PEACEFUL PURPOSES

WITH THE OUTER SPACE TREATY

NASA & US DEPARTMENT OF STATE

COSIMOREPORTS

NEW YORK

The Artemis Accords: Principles for Cooperation in the Civil Exploration, and Use of the Moon, Mars, Comets, and Asteroids for Peaceful Purposes—with the Outer Space Treaty
Originally published in 1967 by the United Nations, published in 2020 by NASA and the US Department of State
This edition published by Cosimo Reports in 2023

ISBN: 978-1-64679-903-9

This edition is a replica of a rare classic. As such, it is possible that some of the text might be blurred or of reduced print quality. Thank you for your understanding, and we wish you a pleasant reading experience.

Ordering Information:
Cosimo publications are available at online bookstores. They may also be purchased for educational, business, or promotional use:
Bulk orders: Special discounts are available on bulk orders for reading groups, organizations, businesses, and others.
Custom-label orders: We offer selected books with your customized cover or logo of choice.

For more information, contact us at www.cosimobooks.com

The Artemis Accords

PRINCIPLES FOR COOPERATION IN THE CIVIL EXPLORATION, AND USE OF THE MOON, MARS, COMETS, AND ASTEROIDS FOR PEACEFUL PURPOSES

WITH THE OUTER SPACE TREATY

NASA & US DEPARTMENT OF STATE

COSIMOREPORTS

NEW YORK

The Artemis Accords: Principles for Cooperation in the Civil Exploration, and Use of the Moon, Mars, Comets, and Asteroids for Peaceful Purposes—with the Outer Space Treaty
Originally published in 1967 by the United Nations, published in 2020 by NASA and the US Department of State
This edition published by Cosimo Reports in 2023

ISBN: 978-1-64679-903-9

This edition is a replica of a rare classic. As such, it is possible that some of the text might be blurred or of reduced print quality. Thank you for your understanding, and we wish you a pleasant reading experience.

Cosimo aims to publish books that inspire, inform, and engage readers worldwide. We use innovative print-on-demand technology that enables books to be printed based on specific customer needs. This approach eliminates an artificial scarcity of publications and allows us to distribute books in the most efficient and environmentally sustainable manner. Cosimo also works with printers and paper manufacturers who practice and encourage sustainable forest management, using paper that has been certified by the FSC, SFI, and PEFC whenever possible.

Ordering Information:
Cosimo publications are available at online bookstores. They may also be purchased for educational, business, or promotional use:
Bulk orders: Special discounts are available on bulk orders for reading groups, organizations, businesses, and others.
Custom-label orders: We offer selected books with your customized cover or logo of choice.

For more information, contact us at www.cosimobooks.com

TABLE OF CONTENTS

I. THE ARTEMIS ACCORDS

THE ARTEMIS ACCORDS

PRINCIPLES FOR COOPERATION IN THE CIVIL EXPLORATION AND

USE OF THE MOON, MARS, COMETS, AND ASTEROIDS

FOR PEACEFUL PURPOSES

TABLE OF CONTENTS

Page Number

i

The Signatories to these Accords;

RECOGNIZING their mutual interest in the exploration and use of outer space for peaceful purposes, and **UNDERSCORING** the continuing importance of existing bilateral space cooperation agreements;

NOTING the benefit for all humankind to be gained from cooperating in the peaceful use of outer space;

USHERING in a new era of exploration, more than 50 years after the historic Apollo 11 Moon landing and more than 20 years after the establishment of a continuous human presence aboard the International Space Station;

SHARING a common spirit and the ambition that the next steps of humanity's journey in space inspire current and future generations to explore the Moon, Mars, and beyond;

BUILDING upon the legacy of the Apollo program, which benefited all of humankind, the Artemis program will land the first woman and next man on the surface of the Moon and establish, together with international and commercial partners, the sustainable human exploration of the solar system;

CONSIDERING the necessity of greater coordination and cooperation between and among established and emerging actors in space;

RECOGNIZING the global benefits of space exploration and commerce;

ACKNOWLEDGING a collective interest in preserving outer space heritage;

AFFIRMING the importance of compliance with the *Treaty on Principles Governing the Activities of States in the Exploration and Use of Outer Space, Including the Moon and Other Celestial Bodies*, opened for signature on January 27, 1967 ("Outer Space Treaty") as well as the *Agreement on the Rescue of Astronauts, the Return of Astronauts and the Return of Objects Launched into Outer Space*, opened for signature on April 22, 1968 ("Rescue and Return Agreement"), the *Convention on International Liability for Damage Caused by Space Objects*, opened for signature on March 29, 1972 ("Liability Convention"), and the *Convention on Registration of Objects Launched into Outer Space,* opened for signature on January 14, 1975 ("Registration Convention"); as well as the benefits of coordination via multilateral forums, such as the United Nations Committee on the Peaceful Uses of Outer Space ("COPUOS"), to further efforts toward a global consensus on critical issues regarding space exploration and use; and

DESIRING to implement the provisions of the Outer Space Treaty and other relevant international instruments and thereby establish a political understanding regarding mutually beneficial practices for the future exploration and use of outer space, with a focus on activities conducted in support of the Artemis Program;

COMMIT to the following principles:

1

SECTION 1 - PURPOSE AND SCOPE

The purpose of these Accords is to establish a common vision via a practical set of principles, guidelines, and best practices to enhance the governance of the civil exploration and use of outer space with the intention of advancing the Artemis Program. Adherence to a practical set of principles, guidelines, and best practices in carrying out activities in outer space is intended to increase the safety of operations, reduce uncertainty, and promote the sustainable and beneficial use of space for all humankind. The Accords represent a political commitment to the principles described herein, many of which provide for operational implementation of important obligations contained in the Outer Space Treaty and other instruments.

The principles set out in these Accords are intended to apply to civil space activities conducted by the civil space agencies of each Signatory. These activities may take place on the Moon, Mars, comets, and asteroids, including their surfaces and subsurfaces, as well as in orbit of the Moon or Mars, in the Lagrangian points for the Earth-Moon system, and in transit between these celestial bodies and locations. The Signatories intend to implement the principles set out in these Accords through their own activities by taking, as appropriate, measures such as mission planning and contractual mechanisms with entities acting on their behalf.

SECTION 2 - IMPLEMENTATION

1. Cooperative activities regarding the exploration and use of outer space may be implemented through appropriate instruments, such as Memoranda of Understanding, Implementing Arrangements under existing Government-to-Government Agreements, Agency-to-Agency arrangements, or other instruments. These instruments should reference these Accords and include appropriate provisions for implementing the principles contained in these Accords.

    (a) In the instruments described in this Section, the Signatories or their subordinate agencies should describe the nature, scope, and objectives of the civil cooperative activity;

    (b) The Signatories' bilateral instruments referred to above are expected to contain other provisions necessary to conduct such cooperation, including those related to liability, intellectual property, and the transfer of goods and technical data;

    (c) All cooperative activities should be carried out in accordance with the legal obligations applicable to each Signatory; and

    (d) Each Signatory commits to taking appropriate steps to ensure that entities acting on its behalf comply with the principles of these Accords.

SECTION 3 – PEACEFUL PURPOSES

The Signatories affirm that cooperative activities under these Accords should be exclusively for peaceful purposes and in accordance with relevant international law.

SECTION 4 – TRANSPARENCY

The Signatories are committed to transparency in the broad dissemination of information regarding their national space policies and space exploration plans in accordance with their national rules and regulations.

The Signatories plan to share scientific information resulting from their activities pursuant to these Accords with the public and the international scientific community on a good-faith basis, and consistent with Article XI of the Outer Space Treaty.

SECTION 5 – INTEROPERABILITY

The Signatories recognize that the development of interoperable and common exploration infrastructure and standards, including but not limited to fuel storage and delivery systems, landing structures, communications systems, and power systems, will enhance space-based exploration, scientific discovery, and commercial utilization. The Signatories commit to use reasonable efforts to utilize current interoperability standards for space-based infrastructure, to establish such standards when current standards do not exist or are inadequate, and to follow such standards.

SECTION 6 – EMERGENCY ASSISTANCE

The Signatories commit to taking all reasonable efforts to render necessary assistance to personnel in outer space who are in distress, and acknowledge their obligations under the Rescue and Return Agreement.

SECTION 7 – REGISTRATION OF SPACE OBJECTS

For cooperative activities under these Accords, the Signatories commit to determine which of them should register any relevant space object in accordance with the Registration Convention. For activities involving a non-Party to the Registration Convention, the Signatories intend to cooperate to consult with that non-Party to determine the appropriate means of registration.

SECTION 8 – RELEASE OF SCIENTIFIC DATA

1. The Signatories retain the right to communicate and release information to the public regarding their own activities. The Signatories intend to coordinate with each other in advance regarding the public release of information that relates to the other Signatories' activities under these Accords in order to provide appropriate protection for any proprietary and/or export-controlled information.

2. The Signatories are committed to the open sharing of scientific data. The Signatories plan to make the scientific results obtained from cooperative activities under these Accords available to the public and the international scientific community, as appropriate, in a timely manner.

3. The commitment to openly share scientific data is not intended to apply to private sector operations unless such operations are being conducted on behalf of a Signatory to the Accords.

SECTION 9 – PRESERVING OUTER SPACE HERITAGE

1. The Signatories intend to preserve outer space heritage, which they consider to comprise historically significant human or robotic landing sites, artifacts, spacecraft, and other evidence of activity on celestial bodies in accordance with mutually developed standards and practices.

2. The Signatories intend to use their experience under the Accords to contribute to multilateral efforts to further develop international practices and rules applicable to preserving outer space heritage. ·

SECTION 10 – SPACE RESOURCES

1. The Signatories note that the utilization of space resources can benefit humankind by providing critical support for safe and sustainable operations.

2. The Signatories emphasize that the extraction and utilization of space resources, including any recovery from the surface or subsurface of the Moon, Mars, comets, or asteroids, should be executed in a manner that complies with the Outer Space Treaty and in support of safe and sustainable space activities. The Signatories affirm that the extraction of space resources does not inherently constitute national appropriation under Article II of the Outer Space Treaty, and that contracts and other legal instruments relating to space resources should be consistent with that Treaty.

3. The Signatories commit to informing the Secretary-General of the United Nations as well as the public and the international scientific community of their space resource extraction activities in accordance with the Outer Space Treaty.

4. The Signatories intend to use their experience under the Accords to contribute to multilateral efforts to further develop international practices and rules applicable to the extraction and utilization of space resources, including through ongoing efforts at the COPUOS.

SECTION 11 – DECONFLICTION OF SPACE ACTIVITIES

1. The Signatories acknowledge and reaffirm their commitment to the Outer Space Treaty, including those provisions relating to due regard and harmful interference.

2. The Signatories affirm that the exploration and use of outer space should be conducted with due consideration to the United Nations Guidelines for the Long-term Sustainability of Outer Space Activities adopted by the COPUOS in 2019, with appropriate changes to reflect the nature of operations beyond low-Earth orbit.

3. Consistent with Article IX of the Outer Space Treaty, a Signatory authorizing an activity under these Accords commits to respect the principle of due regard. A Signatory to these Accords with reason to believe that it may suffer, or has suffered, harmful interference, may request consultations with a Signatory or any other Party to the Outer Space Treaty authorizing the activity.

4. The Signatories commit to seek to refrain from any intentional actions that may create harmful interference with each other's use of outer space in their activities under these Accords.

5. The Signatories commit to provide each other with necessary information regarding the location and nature of space-based activities under these Accords if a Signatory has reason to believe that the other Signatories' activities may result in harmful interference with or pose a safety hazard to its space-based activities.

6. The Signatories intend to use their experience under the Accords to contribute to multilateral efforts to further develop international practices, criteria, and rules applicable to the definition and determination of safety zones and harmful interference.

7. In order to implement their obligations under the Outer Space Treaty, the Signatories intend to provide notification of their activities and commit to coordinating with any relevant actor to avoid harmful interference. The area wherein this notification and coordination will be implemented to avoid harmful interference is referred to as a 'safety zone'. A safety zone should be the area in which nominal operations of a relevant activity or an anomalous event could reasonably cause harmful interference. The Signatories intend to observe the following principles related to safety zones:

   (a) The size and scope of the safety zone, as well as the notice and coordination, should reflect the nature of the operations being conducted and the environment that such operations are conducted in;
   (b) The size and scope of the safety zone should be determined in a reasonable manner

leveraging commonly accepted scientific and engineering principles;

(c) The nature and existence of safety zones is expected to change over time reflecting the status of the relevant operation. If the nature of an operation changes, the operating Signatory should alter the size and scope of the corresponding safety zone as appropriate. Safety zones will ultimately be temporary, ending when the relevant operation ceases; and

(d) The Signatories should promptly notify each other as well as the Secretary-General of the United Nations of the establishment, alteration, or end of any safety zone, consistent with Article XI of the Outer Space Treaty.

8. The Signatory maintaining a safety zone commits, upon request, to provide any Signatory with the basis for the area in accordance with the national rules and regulations applicable to each Signatory.

9. The Signatory establishing, maintaining, or ending a safety zone should do so in a manner that protects public and private personnel, equipment, and operations from harmful interference. The Signatories should, as appropriate, make relevant information regarding such safety zones, including the extent and general nature of operations taking place within them, available to the public as soon as practicable and feasible, while taking into account appropriate protections for proprietary and export-controlled information.

10. The Signatories commit to respect reasonable safety zones to avoid harmful interference with operations under these Accords, including by providing prior notification to and coordinating with each other before conducting operations in a safety zone established pursuant to these Accords.

11. The Signatories commit to use safety zones, which will be expected to change, evolve, or end based on the status of the specific activity, in a manner that encourages scientific discovery and technology demonstration, as well as the safe and efficient extraction and utilization of space resources in support of sustainable space exploration and other operations. The Signatories commit to respect the principle of free access to all areas of celestial bodies and all other provisions of the Outer Space Treaty in their use of safety zones. The Signatories further commit to adjust their usage of safety zones over time based on mutual experiences and consultations with each other and the international community.

SECTION 12 - ORBITAL DEBRIS

1. The Signatories commit to plan for the mitigation of orbital debris, including the safe, timely, and efficient passivation and disposal of spacecraft at the end of their missions, when appropriate, as part of their mission planning process. In the case of cooperative missions, such plans should explicitly include which Signatory has the primary responsibility for the end-of-mission planning and implementation.

2. The Signatories commit to limit, to the extent practicable, the generation of new, long-lived harmful debris released through normal operations, break-up in operational or post-mission

phases, and accidents and conjunctions, by taking appropriate measures such as the selection of safe flight profiles and operational configurations as well as post-mission disposal of space structures.

SECTION 13 – FINAL PROVISIONS

1. Building on any consultative mechanisms in preexisting arrangements as appropriate, the Signatories commit to periodically consult to review the implementation of the principles in these Accords, and to exchange views on potential areas of future cooperation.

2. The Government of the United States of America will maintain the original text of these Accords and transmit to the Secretary-General of the United Nations a copy of these Accords, which is not eligible for registration under Article 102 of the Charter of the United Nations, with a view to its circulation to all the members of the Organization as an official document of the United Nations.

3. After October 13, 2020, any State seeking to become a Signatory to these Accords may submit its signature to the Government of the United States for addition to this text.

Adopted on October 13, 2020, in the English language.

FOR AUSTRALIA

Dr Megan Clark AC
Head, Australian Space Agency

Date: 13 October 2020

FOR CANADA:

Lisa Campbell
President
Canadian Space Agency

Date: 13.10.20

FOR REPUBLIC OF ITALY:

On. Riccardo Fraccaro
Undersecretary of State at the Presidency
of the Council of Ministers

Date: _____ 1 3 OTT. 2020

FOR JAPAN:

INOUE Shinji
Minister of State for Space Policy

Date: 2020/10/13

FOR JAPAN:

萩生田 光一

HAGIUDA Koichi
Minister of Education, Culture, Sports,
Science and Technology

Date: 2020/10/13

FOR THE GRAND DUCHY OF
LUXEMBOURG

Franz Fayot
Minister of the Economy

Date: October 13, 2020

FOR THE UNITED ARAB EMIRATES:

Her Excellency Sarah bint Yousef Al Amiri
Minister of State for Advanced Technologies
Chairwoman of UAE Space Agency

Date: 13. 10. 2020

SIGNED

FOR THE UK SPACE AGENCY
ON BEHALF OF THE GOVERNMENT OF THE
UNITED KINGDOM:

Dr Graham Turnock
Chief Executive

Place: 71st International Astronautical Congress

Date: 13th October 2020_____

FOR THE UNITED STATES OF AMERICA:

James F. Bridenstine
Administrator
National Aeronautics and Space Administration

Date: 10/13/20

II. THE OUTER SPACE TREATY

THE OUTER SPACE TREATY

The *Outer Space Treaty* represents the basic legal framework of international space law and, among its principles, it bars States Parties to the Treaty from placing nuclear weapons or any other weapons of mass destruction in orbit of Earth, installing them on the Moon or any other celestial body, or to otherwise station them in outer space. It exclusively limits the use of the Moon and other celestial bodies to peaceful purposes and expressly prohibits their use for testing weapons of any kind, conducting military manoeuvers, or establishing military bases, installations, and fortifications (Art.IV). Moreover, it explicitly forbids any government from claiming a celestial resource such as the Moon or a planet. Art. II of the Treaty states, in fact, that "outer space, including the moon and other celestial bodies, is not subject to national appropriation by claim of sovereignty, by means of use or occupation, or by any other means."

Treaty on principles governing the activities of states in the exploration and use of outer space, including the moon and other celestial bodies.

The States Parties to this Treaty,

Inspired by the great prospects opening up before mankind as a result of man's entry into outer space,

Recognizing the common interest of all mankind in the progress of the exploration and use of outer space for peaceful purposes,

Believing that the exploration and use of outer space should be carried on for the benefit of all peoples irrespective of the degree of their economic or scientific development,

Desiring to contribute to broad international co-operation in the scientific as well as the legal aspects of the exploration and use of outer space for peaceful purposes,

Believing that such co-operation will contribute to the development of mutual understanding and to the strengthening of friendly relations between States and peoples,

Recalling resolution 1962 (XVIII), entitled "Declaration of Legal Principles Governing the Activities of States in the Exploration and Use of Outer Space", which was adopted unanimously by the United Nations General Assembly on 13 December 1963,

Recalling resolution 1884 (XVIII), calling upon States to refrain from placing in orbit around the earth any objects carrying nuclear weapons or any other kinds of weapons of mass destruction or from installing such weapons on celestial bodies, which was adopted unanimously by the United Nations General Assembly on 17 October 1963,

Taking account of United Nations General Assembly resolution 110 (II) of 3 November 1947, which condemned propaganda designed or likely to provoke or encourage any threat to the peace, breach of the peace or act of aggression, and considering that the aforementioned resolution is applicable to outer space,

Convinced that a Treaty on Principles Governing the Activities of States in the Exploration and Use of Outer Space, including the Moon and Other Celestial Bodies, will further the Purposes and Principles of the Charter of the United Nations,

Have agreed on the following:

Article I

The exploration and use of outer space, including the moon and other celestial bodies, shall be carried out for the benefit and in the interests of all countries, irrespective of their degree of economic or scientific development, and shall be the province of all mankind.

Outer space, including the moon and other celestial bodies, shall be free for exploration and use by all States without discrimination of any kind, on a basis of equality and in accordance with international law, and there shall be free access to all areas of celestial bodies.

There shall be freedom of scientific investigation in outer space, including the moon and other celestial bodies, and States shall facilitate and encourage international co-operation in such investigation.

Article II

Outer space, including the moon and other celestial bodies, is not subject to national appropriation by claim of sovereignty, by means of use or occupation, or by any other means.

Article III

States Parties to the Treaty shall carry on activities in the exploration and use of outer space, including the moon and other celestial bodies, in accordance with international law, including the Charter of the United Nations, in the interest of maintaining international peace and security and promoting international co-operation and understanding.

Article IV

States Parties to the Treaty undertake not to place in orbit around the earth any objects carrying nuclear weapons or any other kinds of weapons of mass destruction, install such weapons on celestial bodies, or station such weapons in outer space in any other manner.

The moon and other celestial bodies shall be used by all States Parties to the Treaty exclusively for peaceful purposes. The establishment of military bases, installations and fortifications, the testing of any type of weapons and the conduct of military manoeuvres on celestial bodies shall be forbidden. The use of military personnel for scientific research or for any other peaceful purposes shall not be prohibited. The use of any equipment or facility necessary for peaceful exploration of the moon and other celestial bodies shall also not be prohibited.

Article V

States Parties to the Treaty shall regard astronauts as envoys of mankind in outer space and shall render to them all possible assistance in the event of accident, distress, or emergency landing on the territory of another State Party or on the high seas. When astronauts make such a landing, they shall be safely and promptly returned to the State of registry of their space vehicle.

In carrying on activities in outer space and on celestial

bodies, the astronauts of one State Party shall render all possible assistance to the astronauts of other States Parties.

States Parties to the Treaty shall immediately inform the other States Parties to the Treaty or the Secretary-General of the United Nations of any phenomena they discover in outer space, including the Moon and other celestial bodies, which could constitute a danger to the life or health of astronauts.

Article VI

States Parties to the Treaty shall bear international responsibility for national activities in outer space, including the Moon and other celestial bodies, whether such activities are carried on by governmental agencies or by non-governmental entities, and for assuring that national activities are carried out in conformity with the provisions set forth in the present Treaty. The activities of non-governmental entities in outer space, including the Moon and other celestial bodies, shall require authorization and continuing supervision by the appropriate State Party to the Treaty. When activities are carried on in outer space, including the moon and other celestial bodies, by an international organization, responsibility for compliance with this Treaty shall be borne both by the international organization and by the States Parties to the Treaty participating in such organization.

Article VII

Each State Party to the Treaty that launches or procures the launching of an object into outer space, including the Moon and other celestial bodies, and each State Party from whose territory or facility an object is launched, is internationally liable for damage to another State Party to the Treaty or to its natural or juridical persons by such object or its component parts on the Earth, in air space or in outer space, including the Moon and other celestial bodies.

Article VII

A State Party to the Treaty on whose registry an object launched into outer space is carried shall retain jurisdiction and control over such object, and over any personnel thereof, while in outer space or on a celestial body. Ownership of objects launched into outer space, including objects landed or constructed on a celestial body, and of their component parts, is not affected by their presence in outer space or on a celestial body or by their return to the Earth. Such objects or component parts found beyond the limits of the State Party of the Treaty on whose registry they are carried shall be returned to that State Party, which shall, upon request, furnish identifying data prior to their return.

Article IX

In the exploration and use of outer space, including the Moon and other celestial bodies, States Parties to the Treaty shall be guided by the principle of co-operation and mutual assistance and shall conduct all their activities in outer space, including the moon and other celestial bodies, with due regard to the corresponding interests of all other States Parties to the Treaty. States Parties to the Treaty shall pursue studies of outer space, including the Moon and other celestial bodies, and conduct exploration of them so as to avoid their harmful contamination and also adverse changes in the environment of the Earth resulting from the introduction of extraterrestrial matter and, where necessary, shall adopt appropriate measures for this purpose. If a State Party to the Treaty has reason to believe that an activity or experiment planned by it or its nationals in outer space, including the Moon and other celestial bodies, would cause potentially harmful interference with activities of other States Parties in the peaceful exploration and use of outer space, including the moon and other celestial bodies, it shall undertake appropriate international consultations before proceeding with any such activity or experiment. A State Party to the Treaty which has reason to believe that an activity or experiment planned by another State Party in outer space, including the Moon and other celestial bodies, would cause potentially harmful interference with activities in the peaceful exploration and use of outer space, including the moon and other celestial bodies, may request consultation concerning the activity or experiment.

Article X

In order to promote international co-operation in the exploration and use of outer space, including the Moon and other celestial bodies, in conformity with the purposes of this Treaty, the States Parties to the Treaty shall consider on a basis of equality any requests by other States Parties to the Treaty to be afforded an opportunity to observe the flight of space objects launched by those States.

The nature of such an opportunity for observation and the conditions under which it could be afforded shall be determined by agreement between the States concerned.

Article XI

In order to promote international co-operation in the peaceful exploration and use of outer space, States Parties to the Treaty conducting activities in outer space, including the Moon and other celestial bodies, agree to inform the Secretary-General of the United Nations as well as the public and the international scientific community, to the greatest extent feasible and practicable, of the nature, conduct, locations and results of such activities. On receiving the said information, the Secretary-General of the United Nations should be prepared to disseminate it immediately and effectively.

Article XII

All stations, installations, equipment and space vehicles on the moon and other celestial bodies shall be open to representatives of other States Parties to the Treaty on a basis of reciprocity. Such representatives shall give reasonable advance notice of a projected visit, in order that appropriate consultations may be held and that maximum precautions may be taken to assure safety and to avoid interference with normal operations in the facility to be visited.

Article XIII

The provisions of this Treaty shall apply to the activities of States Parties to the Treaty in the exploration and use of outer space, including the Moon and other celestial bodies, whether such activities are carried on by a single State Party to the Treaty or jointly with other States, including cases where they are carried on within the framework of international inter-governmental organizations.

Any practical questions arising in connexion with activities carried on by international inter-governmental organizations in the exploration and use of outer space, including the moon and other celestial bodies, shall be resolved by the States Parties to the Treaty either with the appropriate international organization or with one or more States members of that international organization, which are Parties to this Treaty.

Article XIV

1. This Treaty shall be open to all States for signature. Any State which does not sign this Treaty before its entry into force in accordance with paragraph 3 of this Article may accede to it at any time.

2. This Treaty shall be subject to ratification by signatory States. Instruments of ratification and instruments of accession shall be deposited with the Governments of the United Kingdom of Great Britain and Northern Ireland, the Union of Soviet Socialist Republics and the United States of America, which are hereby designated the Depositary Governments.

3. This Treaty shall enter into force upon the deposit of instruments of ratification by five Governments including the Governments designated as Depositary Governments under this Treaty.

4. For States whose instruments of ratification or accession are deposited subsequent to the entry into force of this Treaty, it shall enter into force on the date of the deposit of their instruments of ratification or accession.

5. The Depositary Governments shall promptly inform all signatory and acceding States of the date of each signature, the date of deposit of each instrument of ratification of and accession to this Treaty, the date of its entry into force and other notices.

6. This Treaty shall be registered by the Depositary Governments pursuant to Article 102 of the Charter of the United Nations.

Article XV

Any State Party to the Treaty may propose amendments to this Treaty. Amendments shall enter into force for each State Party to the Treaty accepting the amendments upon their acceptance by a majority of the States Parties to the Treaty and thereafter for each remaining State Party to the Treaty on the date of acceptance by it.

Article XVI

Any State Party to the Treaty may give notice of its withdrawal from the Treaty one year after its entry into force by written notification to the Depositary Governments. Such withdrawal shall take effect one year from the date of receipt of this notification.

Article XVII

This Treaty, of which the Chinese, English, French, Russian and Spanish texts are equally authentic, shall be deposited in the archives of the Depositary Governments. Duly certified copies of this Treaty shall be transmitted by the Depositary

Governments to the Governments of the signatory and acceding States.

In witness whereof the undersigned, duly authorised, have signed this Treaty.

Done at the cities of London, Moscow and Washington, the 27th day of January, one thousand nine hundred and sixty-seven.

A SELECTION OF COSIMO REPORTS ON US SPACE POLICY

National Security Space Strategy—Unclassified Summary	ISBN 978-1-64679-905-3 (Cosimo Reports, 2011)
Defense Space Strategy—Summary	ISBN 978-1-64679-904-6 (Cosimo Reports, 2020)
National Space Policy of the United States of America	ISBN 978-1-64679-906-0 (Cosimo Reports, 2020)
Spacepower—Doctrine for Space Forces	ISBN 978-1-64679-908-4 (Cosimo Reports, 2020)
The Artemis Accords & The Outer Space Treaty	ISBN 978-1-64679-903-9 (Cosimo Reports, 2020)
Space Operations—Doctrine for Space Forces	ISBN 978-1-64679-907-7 (Cosimo Reports, 2022)

And visit cosimobooks.com for many more reports that affect your world, from global trends to the economy, and from health to geopolitics.

COSIMO is a specialty publisher for independent authors, not-for-profit organizations, and innovative businesses, dedicated to publishing books that inspire, inform, and engage readers around the world.

Our mission is to create a smart and sustainable society by connecting people with valuable ideas. We offer authors and organizations full publishing support, while using the newest technologies to present their works in the most timely and effective way.

COSIMO BOOKS offers fine books that inspire, inform and engage readers on a variety of subjects, including personal development, socially responsible business, economics and public affairs.

COSIMO CLASSICS brings to life unique and rare classics, representing a wide range of subjects that include Business, Economics, History, Personal Development, Philosophy, Religion & Spirituality, and much more!

COSIMO REPORTS publishes reports that affect your world, from global trends to the economy, and from health to geopolitics.

COSIMO B2B offers custom editions for historical societies, museums, companies and other organizations interested in offering classic books to their audiences, customized with their own logo and message. **COSIMO B2B** also offers publishing services to organizations, such as media firms, think tanks, conference organizers and others who could benefit from having their own imprint.

www.ingramcontent.com/pod-product-compliance
Lightning Source LLC
LaVergne TN
LVHW010149020426
835378LV00011B/1025